DAIRY PRODUCTS

FARM TO MARKET

Jason Cooper

Rourke Publications, Inc.
Vero Beach, Florida 32964

© 1997 Rourke Publications, Inc.

All rights reserved. No part of this book may be reproduced or utilized in any form or by any means, electronic or mechanical including photocopying, recording or by any information storage and retrieval system without permission in writing from the publisher.

Edited by Pamela J.P. Schroeder

PHOTO CREDITS
All photos © Lynn M. Stone

Library of Congress Cataloging-in-Publication Data
Cooper, Jason, 1942-
    Dairy products / by Jason Cooper.
        p. cm. — (Farm to market)
    Summary: Describes where North American dairy cattle are raised, what dairy products are, and how they are processed and marketed.
    ISBN 0-86625-619-9
    1. Dairying—Juvenile literature. 2. Dairy products—Juvenile literature. 3. Dairying—North America—Juvenile literature. 4. Dairy products—North America—Juvenile literature.
[1. Dairying. 2. Dairy products.] I. Title. II. Series: Cooper, Jason, 1942-   Farm to market.
SF239.5.C66   1997
641.3'7—dc21   97-13231
                                                CIP
                                                 AC

Printed in the USA

TABLE OF CONTENTS

Dairy Products	5
Dairy Cattle	6
Where Dairy Cattle Are Raised	9
How Dairy Cattle Are Raised	11
Milking Dairy Cattle	14
Farm to Factory	16
Processing Milk	19
Making Dairy Products	20
Dairy Products as Food	22
Glossary	23
Index	24

DAIRY PRODUCTS

Milk and foods made from milk are **dairy** (DARE ee) products. A dairy is a place where milk is stored, collected, or sold.

Most dairy products in North America are made from the milk of cattle. Some goat milk and cheese is also produced in North America.

In addition to milk and cream, dairy products include foods such as butter, ice cream, sherbet, yogurt, buttermilk, and cheese.

An Ayrshire dairy cow stands in a Vermont pasture. Today, dairy cows in the United States produce three times as much milk per cow as in 1940.

DAIRY CATTLE

Dairy cattle, often called cows, are females raised to produce milk on farms.

North American farmers raise several **breeds** (BREEDZ), or types, of dairy cattle. About nine of every 10 dairy cows in North America are Holsteins. Most Holstein cattle are black and white.

Holsteins give more milk than other breeds. Smaller breeds, like Guernseys, Jerseys, Ayrshires, and brown Swiss give richer milk than Holsteins. Rich milk has more cream in it.

A Holstein cow seems to enjoy chewing her cud on an autumn afternoon. Holsteins are big cows. They produce more milk, on average, than any other breed.

WHERE DAIRY CATTLE ARE RAISED

Dairy cattle are raised all across North America. California produces the most milk, followed by Wisconsin, New York, Pennsylvania, and Minnesota. Alaska produces less milk than any other state.

Wisconsin has more milk cows than any state, about 1,500,000. Wisconsin is known as "America's Dairyland," even though California cows produce more milk.

At many dairy farms, farmers keep cows in barns or barnyards. Cows are hardy animals. They do well in Florida's hot summers and Minnesota's cold winters.

More dairy cows are in Wisconsin than any other state. California, however, produces more milk than Wisconsin.

HOW DAIRY CATTLE ARE RAISED

Some farms still let their dairy herds go into pastures during warm months. The cows graze on grasses. Even cows on pastures, however, get more food when they come back to the barn. By watching what cows eat, farmers can make them give more milk.

Each cow gives a different amount of milk. The breed is important, but so too is feed and climate. Cows in mild climates, like California's, produce more milk per cow than cows in colder climates.

The average cow in the United States gives about 20,000 pounds (9,090 kilograms) of milk each year.

Pastures are an important source of food for many dairy cows during the warm months.

Workers at a Wisconsin cheese factory separate blocks of cheddar cheese as it hardens.

Female, or heifer, calves become the adult cows that provide milk for dairy products. Bull calves are usually sold for veal meat.

MILKING DAIRY CATTLE

Pregnant cows and cows with calves produce milk. The milk is in the cow's **udder** (UH der), or milk bag.

At least twice each day, a farmer milks the cows in a barn. The farmer attaches an electric milking machine to each cow's udder. The machine drains milk from the udder. It also pumps the milk through pipes into a cooling tank in the barn.

The average Holstein cow in the United States gives about 70 pounds (32 kg) of milk each day.

A dairy farmer puts a milking machine on her Guernsey's udder. Cows are generally milked twice each day.

FARM TO FACTORY

Milk is transported by truck from dairy farms to milk **processing** (PRAH sess ing) plants. Milk trucks hold milk in large steel tanks.

The milk truck driver tests the milk at each farm stop before pumping it into the truck. The tests tell whether the milk is free of disease. They also tell how rich the milk is. Farmers are paid for how much milk they have and how rich it is.

When the truck reaches a processing plant, workers there test the milk again.

Milk from this Wisconsin dairy farm is being shipped in a stainless steel tank truck to a milk-processing plant.

PROCESSING MILK

Milk processing plants take milk through many steps. First, the plant separates the cream from the milk. Next, the milk is **pasteurized** (PAST ur izd), or heated to kill harmful **bacteria** (bak TEER ee ah). We can't see bacteria, but some kinds can cause disease.

After being heated, milk is **homogenized** (huh MAH jen izd) to mix the cream and milk together. Processing plants also add some healthy ingredients to milk, such as vitamin D. The final step is putting milk into bottles or cartons.

Milk is processed into many foods, including the yogurt, butter, cheese, and whole milk you see here.

MAKING DAIRY PRODUCTS

Each dairy product is made from milk that is processed in a special way. Ice cream, for example, is made from milk with sugar, water, and flavor.

Milk that is used for cheese goes through many steps. First, milk is turned into **curd** (KURD). Curd is soft and puddinglike.

Most dairy products are made in large processing plants. Trains and trucks take the products from the plants to supermarkets and other stores.

A dairy plant worker makes cottage cheese from milk curd.

DAIRY PRODUCTS AS FOOD

Milk is mostly water, but it also has many **nutrients** (NU tree ents). Nutrients, such as vitamins and minerals, are what we need for healthy bodies.

Milk is a major source of calcium, a mineral. Calcium helps make strong bones and teeth.

Milk also has large amounts of vitamins A and B-2. In fact, milk has more vitamins, and in greater amounts, than most natural foods.

Whole milk keeps its cream. Cream is the source of fat in milk. Skim milk is milk without cream. It has almost no fat.

Glossary

bacteria (bak TEER ee ah) — a group of tiny living things that can be seen only with a microscope, both helpful and harmful to people

breed (BREED) — a particular group or type of farm animal within a larger group of very closely related animals, such as a *Holstein* cow among all cows

curd (KURD) — when milk is processed in a special way; a major part of cheese that is lumpy and protein-rich

dairy (DARE ee) — a place where milk is processed for sale

homogenized (huh MAH jen izd) — the process in which milk and cream are mixed, making "whole" milk

nutrient (NU tree ent) — any of several "good" substances that the body needs for health, growth, and energy; vitamins and minerals

pasteurized (PAST ur izd) — a heating process that kills germs in milk

processing (PRAH sess ing) — the steps to prepare fresh fruit, meat, or milk for the market

udder (UH der) — the milk gland, or sack, of a cow

INDEX

bacteria   19
barns   9, 11, 14
breeds (cattle)   6
calcium   22
California   9
cattle   5, 6, 9
cheese   5, 20
cows   6, 9, 11, 14
curd   20
dairy   5, 6, 9, 11, 16
dairy products   5, 20
disease   16, 19
farmers   6, 11, 14, 16
farms   6, 9, 11, 16
fat   22
Holsteins   6, 14
ice cream   5, 20

milk   5, 6, 9, 11, 14, 16, 19, 20, 22
  skim   22
  whole   22
milking machine   14
nutrients   22
plants, processing   16, 19, 20
supermarkets   20
trucks   16, 20
udder   14
vitamins   22
Wisconsin   9